TECHNICAL REPORT

Should the Increase in Military Pay Be Slowed?

James Hosek • Beth J. Asch • Michael G. Mattock

Prepared for the Office of the Secretary of Defense

NATIONAL DEFENSE RESEARCH INSTITUTE

The research described in this report was prepared for the Office of the Secretary of Defense (OSD). The research was conducted within the RAND National Defense Research Institute, a federally funded research and development center sponsored by OSD, the Joint Staff, the Unified Combatant Commands, the Navy, the Marine Corps, the defense agencies, and the defense Intelligence Community under Contract W74V8H-06-C-0002.

Library of Congress Control Number: 2012944168

ISBN: 978-0-8330-7414-0

Cover photo: The USS Essex and the Military Sealift Command fleet replenishment oiler USNS Pecos steam together during a refueling at sea, February 6, 2008. Photo by U.S. Navy Petty Officer 2nd Class Joshua J. Wahl.

Published 2012 by the RAND Corporation
1776 Main Street, P.O. Box 2138, Santa Monica, CA 90407-2138
1200 South Hayes Street, Arlington, VA 22202-5050
4570 Fifth Avenue, Suite 600, Pittsburgh, PA 15213-2665
RAND URL: http://www.rand.org/
To order RAND documents or to obtain additional information, contact
Distribution Services: Telephone: (310) 451-7002;
Fax: (310) 451-6915; Email: order@rand.org

Preface

The Department of Defense (DoD) is seeking cost savings but wants to ensure that the nation's defense requirements continue to be met. The Office of the Under Secretary of Defense for Personnel and Readiness (OSD P&R) asked RAND to provide an assessment of whether the rate of increase in military pay should be slowed. While slowing the increase in pay could provide cost savings, it is important to consider whether doing so would adversely affect the department's ability to recruit an adequate number of high-quality personnel. This assessment assembles the latest available data on recruiting and retention for the active and reserve components, military pay relative to civilian pay, and current employment conditions in the civilian economy. These data are compared with guidelines that are frequently used in determining the annual adjustment in military pay: (1) recruit quality benchmarks (percentage of high school diploma graduates, percentage scoring in the upper half of the Armed Forces Qualification Test score distribution); (2) an index of civilian wage trends (the Employment Cost Index); and (3) civilian wage distributions for workers comparable in terms of age and education (for which the 70th percentile has been established as the benchmark).

This document should be of interest to audiences concerned with national security and the federal deficit. The document's information is perhaps most relevant to the defense manpower policy community and officials charged with ensuring that the nation meets its defense manpower requirements yet does so cost-effectively. The research was sponsored by the Office of the Deputy Assistant Secretary of Defense for Military Personnel Policy and conducted within the Forces and Resources Policy Center of the RAND National Defense Research Institute, a federally funded research and development center sponsored by the Office of the Secretary of Defense, the Joint Staff, the Unified Combatant Commands, the Navy, the Marine Corps, the defense agencies, and the defense Intelligence Community.

For more information on the RAND Forces and Resources Policy Center, see http://www.rand.org/nsrd/ndri/centers/frp.html or contact the director (contact information is provided on the web page).

Contents

Figures

Tables

Summary

Should the Department of Defense (DoD) slow the increase in military pay in coming years? And by doing so, can it achieve desired budget efficiencies while maintaining a workforce of sufficient size and quality to meet future military objectives? We believe the answer is yes, given a robust climate for recruiting and retention, expected changes in the size of the force, and favorable comparison between military and civilian pay. Therefore, we recommend that the rate of increase in military pay be slowed, and we offer several alternatives for implementation.

As the first half of fiscal year (FY) 2012 draws to a close, recruiting and retention remain in excellent shape. With a few exceptions, the military services' active and reserve components have met or exceeded their recruiting and retention targets since 2009. Quality targets for recruits set by the Office of the Secretary of Defense have also been met and often exceeded. Indeed, overall recruit quality has steadily increased.

More recently, the United States ended the war in Iraq and plans to reduce its presence in Afghanistan over the next two years. This offers an opportunity for DoD to reconsider its force structure needs and, in turn, its future manpower requirements. In January, the Secretary of Defense revealed plans to reduce the size of the force by some 72,000 soldiers and 20,000 Marines over the coming years as part of a broader plan to cut departmental spending. As a consequence, recruiting and retention goals in the Army and Marine Corps are likely to be reduced significantly.

The third factor in our calculus, comparisons between military and civilian pay, have been increasingly favorable for much of the past decade, following pay actions put in place in the early 2000s when military pay lagged behind the civilian sector. Comparisons between military and civilian pay have become an important consideration in establishing pay increases for the military. Such comparisons typically involve two approaches. The first is to compare changes in basic pay (one element of military pay) to the Employment Cost Index (ECI) for wages and salaries in the private sector.

The second is to determine how military pay ranks in the wage distribution for civilian workers who are comparable to military personnel in terms of both education and occupation. When this approach is used, military pay is usually measured by regular military compensation, a more expansive measure of pay that includes basic pay, allowances for subsistence and housing, and the federal tax advantage deriving from the fact that allowances are not taxed. Basic pay comprises about 60 percent of cash compensation; regular military compensation comprises slightly over 90 percent. The remainder consists of special and incentive pay (including bonuses) and other allowances.

Military Pay Raises over the Past Decade

The National Defense Authorization Act of 2000 authorized a 6.8 percent increase in basic pay in fiscal year 2000 and basic pay increases equal to the percentage increase in the ECI plus half a percentage point (0.5) through fiscal year 2006.[1] This legislation responded to three challenges facing DoD: (1) a growing gap between military and civilian pay for some portions of the military workforce, primarily in the mid-grade enlisted force; (2) deteriorating recruiting conditions faced by the Army and Marine Corps; and (3) retention difficulties in technical military occupational specialties in the late 1990s.

Following the terrorist attacks of September 11, 2001, and subsequent U.S. military operations in Iraq and Afghanistan, the basic pay increases—the ECI plus half a percentage point—were continued through FY 2010 as insurance against a decline in either the size or quality of the military workforce. In addition to the higher-than-usual increases in basic pay, the housing allowance was raised in the early part of the decade to cover the full expected cost of off-base housing. Together these pay actions succeeded in increasing basic pay and regular military compensation relative to civilian pay.

So, what do these policy decisions mean for military pay today?

- Basic pay is up 45 percent since 2000, substantially more than the ECI (up 33 percent) and the Consumer Price Index (up 31 percent). As a result, military pay buys a lot more than it used to.
- Basic pay, adjusted for inflation, has increased over the decade for officers and enlisted personnel in every age group and service. Since 2000, basic pay has realized average increases of 13 to 18 percent for enlisted personnel ages 18–27, 15 to 24 percent for enlisted personnel ages 28–37, and 13 percent for officers.
- The increases in regular military compensation, adjusted for inflation, are even more dramatic. Regular military compensation is up an average of 40 percent for enlisted members, with some variation by age group and branch of service, and 25 percent for officers. Growth in regular military compensation outpaced that of basic pay due to the increase in the basic allowance for housing.

While real military pay increased over the last decade, civilian pay did not, dropping 4 to 8 percent between 2000 and 2009 for most age and educational groups. As a result, military pay is now well above the 70th percentile of civilian wage distributions for most enlisted and officer age and education groups. In 2000, the Ninth Quadrennial Review of Military Compensation established that regular military compensation should reach at least the 70th percentile of comparable civilian earnings in order to sustain the size and quality force desired by the military departments. Between 2000 and 2009, regular military compensation for enlisted personnel grew from about the 60th percentile of civilian earnings for high school graduates to close to or even exceeding the 80th percentile, depending on the service; regular military compensation for officers increased from the 70th to the 85th percentile of earnings for civilians with a bachelor's degree.

A number of other factors should also be taken into consideration when comparing military and civilian pay. One of the most important is the cost of health care. Active-duty ser-

[1] For example, if the ECI increase was 3 percent, the authorized basic pay increase would be 3.5 percent.

vice members receive health care at no cost, and costs for their families are quite low. If these service members were employed in the civilian labor market, they would likely be required to contribute to health plan premiums in order to obtain similar coverage. These contributions reduce take-home pay for civilians. In 2000, average annual worker contributions to premiums for single and family coverage were $334 and $1,619, respectively. Moreover, as the cost of health care rises in the civilian sector, the value to service members of not paying health care costs increases. Between 2000 and 2010, health plan premiums increased approximately 150 percent—a growth rate that far exceeded the 31 percent increase in the cost of living over the same period. After incorporating the value of avoiding health care costs, military pay places even higher on the civilian wage distribution.

Furthermore, civilian employment conditions have been poor since early 2008, when the unemployment rate rose precipitously. The unemployment rate increased for all education levels and remains high at about 8.3 percent. Many of the unemployed have been so for long periods (more than 26 weeks), and the number of workers employed part-time has doubled since 2008, due in large part to the poor condition of the labor market. These factors affect the wages civilians can *expect* to receive—an aspect not fully reflected by the ECI, which accounts only for wages paid to employed workers. The distinction is significant: the *expected* weekly wage is 2 to 6 percent less than the average weekly wage of full-time workers. This difference translates to a reduction in earnings of about $25 per week for college graduates and $40–$50 per week for those without a college degree. Accounting for the possibility of unemployment further increases the gap between military and civilian pay.

Implementation Strategies

If DoD elects to move forward with limited pay increases in 2015, it must consider how to implement the changes. We offer three implementation strategies for consideration. The first is a one-time pay increase set at half a percentage point below the ECI. The second is a one-year freeze in basic pay. The third is a series of below-ECI increases, such as ECI minus half a percentage point for four years.

Each approach has its advantages and disadvantages. The first option, a one-time half-percentage-point increase below the ECI, yields lower cost savings than the other options—estimated at about $5 billion over ten years. However, it has the advantage of being a more cautious approach and perhaps more palatable politically. The second and third options are likely to be more costly politically. A pay freeze might be taken as a lack of regard for the sacrifices of service members and their families in the immediate aftermath of a decade-long war in Iraq. A prolonged, four-year period of lower-than-usual basic pay increases could be difficult to sustain, particularly if looming pay reductions create uncertainty for members and possibly reduce morale, which could ultimately affect recruiting and retention decisions. Implementing a pay freeze, which service members would likely perceive as a more severe pay reduction, could exacerbate these responses. On the other hand, the second and third options yield substantially greater savings than the first option and are estimated at about $17 billion over the next ten years.

Because we are unable to rigorously measure and assess the relative importance of these factors, we do not recommend a specific implementation approach. Policymakers will need to weigh the advantages and disadvantages of each option before rendering a considered judgment.

What we can recommend is slowing pay growth relative to the ECI, given current conditions and DoD's stated workforce goals in terms of both quantity and quality. Recruiting and retention are in excellent shape, with all services and components meeting or exceeding recruiting and retention goals. Planned reductions in manpower requirements have been announced by the Secretary of Defense. And military pay compares well to civilian pay, ranking significantly above the 70th percentile benchmark. Given these conditions, it is feasible for DoD to slow the increase in military pay, which would enable savings in military personnel costs while achieving force management goals.

We recognize that slowing the growth in military pay could affect recruiting and retention more than expected and that planned reductions in the size of the force could be reversed or slowed. But in either case, targeted bonuses can be used to offset these effects. Further, with changes in military retirement also under consideration, it would be prudent to identify potential interactions between slowed pay growth and retirement changes and to evaluate the implications of those interactions for sustaining the current force. RAND has such research under way, with results expected in the fall of 2012.

Acknowledgments

We thank our RAND colleagues Arthur Bullock and Craig Martin for their help in processing the military pay and Current Population Survey files. We appreciate the guidance on our compensation research received from Jeri Busch, Director of Military Compensation in the Office of the Under Secretary of Defense for Personnel and Readiness; Vee Penrod, Deputy Assistant Secretary of Defense (Military Personnel Policy); and members of their staffs. We also appreciate the help of Curtis Gilroy, Director of Accession Policy in the Office of the Under Secretary of Defense for Personnel and Readiness, and John Jessup, also of that office, for providing recruiting and retention statistics and answering our questions. We received valuable comments from our reviewers, Michael Hansen and Michael Polich, which helped us improve the report.

Abbreviations

AC	active component
AFQT	Armed Forces Qualification Test
BLS	Bureau of Labor Statistics
CBO	Congressional Budget Office
CPI	Consumer Price Index
CPI-U	Consumer Price Index for urban consumers
CPS	Current Population Survey
DoD	Department of Defense
ECI	Employment Cost Index
FY	fiscal year
HOM	*Handbook of Methods*
HSDG	high school diploma graduate
MPP	Military Personnel Policy
NCS	National Compensation Survey
OSD	Office of the Secretary of Defense
OSD (MPP/AP)	Office of the Secretary of Defense for Military Personnel Policy, Accession Policy Directorate
P&R	Personnel and Readiness
QRMC	Quadrennial Review of Military Compensation
RC	reserve components
RMC	regular military compensation

CHAPTER ONE

Introduction

This report provides information relevant to whether basic pay increases in the next budget cycles should be lower than the increase in the benchmark index, the Employment Cost Index (ECI) for private-sector wage and salary workers.

Motivation for the report comes from two sources. The first is that the services should pay the volunteer force enough to meet its manning requirements but should not overpay, as there are alternative uses for scarce resources. The second is the national priority to decrease the federal debt by means of decreases in federal outlays over the next decade. If it appears possible to decrease basic pay relative to private sector pay without jeopardizing the overarching objective of meeting manning requirements with highly capable personnel, then the basic pay budget can be an element of federal cost saving. In fact, Secretary of Defense Leon E. Panetta, in recent testimony before the Senate Armed Services Committee, announced the intention of the Department of Defense (DoD) to "provide more limited pay raises beginning in 2015, giving troops and their families fair notice and lead time before changes take effect." (Department of Defense, 2012).

The ECI has been written into legislation to serve as the guide for adjusting basic pay. (The current legislation on the use of the ECI appears in Appendix A.) The ECI, as a guide, provides a starting point for the annual discussion over the basic pay adjustment. Congress may choose an adjustment that is equal to, lower, or higher than the change in the ECI. The adjustment depends critically on whether the services are meeting their manning requirements—that is, whether recruiting, retention, and recruit quality are satisfactory.

This report begins with data on recruiting and retention. The data confirm that recruiting and retention are in excellent shape at present. The report then presents information on military and civilian pay. This is accompanied by a discussion of a focal point among policymakers on the level of military pay relative to civilian pay, namely, whether military pay is at or above the 70th percentile of civilian pay for workers with comparable characteristics to those of service members. Like the ECI, the 70th percentile is a point of departure for considering whether military pay is at an adequate level. Charts and tables comparing military pay to civilian wages follow this discussion. A key finding is that constant-dollar civilian pay did not increase during the past decade and actually decreased slightly, whereas basic pay increased, and regular military compensation (RMC) and total pay increased even more: RMC rose well above the 70th percentile of civilian pay and was near or even exceeding the 80th percentile in 2009, depending on the service.

The report also argues that the increase in the unemployment rate since 2008 has in effect decreased the expected civilian wage, so the effective increase in military versus civilian pay is even greater than that shown in our civilian pay trend data (which do not adjust for unemployment).

The report then compares the trends in the ECI with those in the Current Population Survey (CPS) data and suggests why the ECI has increased by more than CPS wages since 2000. Finally, it concludes with a discussion of three alternatives for slowing the increase in basic pay: (1) a one-time increase in basic pay below the ECI, (2) a one-year basic pay freeze, or (3) below-ECI basic pay growth over several years.

CHAPTER TWO

Recruiting and Retention Outcomes, 2005–2011

Overview

Enlisted recruiting and retention are now very good, compared with historical norms and targets:

- All of the services are meeting their active component (AC) "volume" goals, and most are meeting the goals of the reserve components (RC). All services are getting essentially 100 percent high school diploma graduates[1] (HSDG). This exceeds the Office of the Secretary of Defense (OSD) goal of at least 90 percent HSDG.
- Recruit test scores are very favorable: Almost two-thirds of Army recruits are in Armed Forces Qualification Test (AFQT) Categories I–IIIA; the other services show an even greater proportion of recruits scoring in Categories I–IIIA. In 2011, 77 percent of DoD-wide recruits were I–IIIA, well above the OSD goal of at least 60 percent.

Outcomes

Tables 2.1 through 2.4 show the enlisted recruiting and retention outcomes for the active and reserve components from 2005 through 2011 as reported by the Office of Accession Policy, OSD Personnel and Readiness. We focus on recruiting and retention figures for enlisted personnel because they make up about 85 percent of the active duty armed forces. The health of military recruiting and retention is largely determined by trends for the enlisted force. The top part of Table 2.1 shows the percentage of the enlisted accession goal that was met by each service and by DoD overall. Over this period, the accession goal varied over time and across services (not shown). The Navy, Air Force, and Marine Corps met their active duty enlisted accession goals throughout this period. The Army met its goals, too, except for 2005.

The middle and bottom sections of Table 2.1 show how the services performed in terms of two metrics of recruit quality—the percentage of enlisted recruits who scored in the upper half of the AFQT score distribution, denoted as being in AFQT Categories I to IIIA, and the percentage of recruits who are HSDGs. While the services have internal recruit quality targets, OSD sets a goal or benchmark that at least 90 percent of each service's recruits must be HSDG and at least 60 percent in AFQT Categories I to IIIA, as defined in DoD Instruction 1145.01

[1] HSDG is a specific category tracked by military recruiters. It means that an individual completed high school, as evidenced by receiving a diploma, and it helps to distinguish these recruits from others who, say, only attended the 12th year of school but did not complete it or those who "completed" high school by passing the GED.

Table 2.1
Active Component Enlisted Recruiting

	2005	2006	2007	2008	2009	2010	2011
Percentage of Accessions Goal Achieved							
Army	92	101	101	101	108	100	100
Navy	100	100	101	100	100	100	100
Marine Corps	100	100	101	100	100	100	100
Air Force	102	100	100	100	100	100	100
DoD	96	100	100	100	103	100	100
Percentage AFQT Categories I to IIIA							
Army	67	61	61	62	66	64	63
Navy	70	75	73	74	78	83	89
Marine Corps	68	68	65	66	71	72	73
Air Force	80	78	79	79	81	90	99
DoD	70	69	68	68	73	74	77
Percentage HSDG							
Army	87	81	79	83	95	100	99
Navy	96	95	93	94	95	98	99
Marine Corps	96	96	95	96	99	100	100
Air Force	99	99	99	99	99	99	100
DoD	93	91	90	92	96	99	99

SOURCE: OSD (MPP/AP).

(Department of Defense, 2005). As described below, these benchmarks are based on the quality and performance of recruits from the early 1990s. Table 2.1 shows that the Army fell below the 90 percent HSDG benchmark between 2005 and 2008. In recent years, the services have far exceeded the quality benchmarks. In 2010 and 2011, nearly 100 percent of all recruits were HSDG; in 2010, 74 percent of all recruits across DoD scored in the upper half of the AFQT score distribution; and in 2011, 77 percent did so. These recent recruiting results are consistent with past research that finds that increases in unemployment increase the attractiveness of military service and increase the supply of high-quality personnel to the armed forces (Simon and Warner, 2007; Asch et al., 2010). The services have been able to take advantage of the increase in supply by being more selective in the quality of recruits they access.

Table 2.2 shows the degree to which each service met its annual retention goals for enlisted AC personnel between 2005 and 2011, by stage of the enlisted career where retention is defined in terms of reenlistments. That is, each service reports to OSD its annual goal for the number of personnel it seeks to reenlist at each career stage, where the stage is defined by the individual service, and each service also reports its success in reaching that goal. For example, in fiscal year (FY) 2010, the Army sought to reenlist 24,500 soldiers at the end of their initial enlistment term, and 27,436 soldiers at this stage of their career reenlisted, implying that the Army reached 112 percent of its goal.

As seen in Table 2.2, the AC enlisted retention goals were generally met throughout 2005–2011. The low retention figures for the Navy in 2005 and the Air Force in 2008 and 2010 reflect drawdowns under way in those services at that time. While it might seem easier to reach retention goals when the need for personnel is declining, these services still needed to retain personnel in specific skill areas. For example, according to the Congressional Budget

Table 2.2
Active Component Enlisted Retention (percentage of goal)

	2005	2006	2007	2008	2009	2010	2011
Army							
Initial	103	106	117	114	138	112	105
Mid-career	103	100	107	114	112	111	110
Career	128	111	111	113	126	123	117
Navy							
Initial	52	91	98	102	107	134	128
Mid-career	63	99	108	98	108	131	116
Career	85	111	109	103	108	134	142
Marine Corps							
Initial	103	102	92	87	109	102	101
Career	138	116	129	104	107	107	129
Air Force							
Initial	N/A	113	99	64	101	93	105
Mid-career	N/A	114	94	84	99.8	107	106
Career	N/A	109	95	79	98.4	98	96

SOURCE: OSD (MPP/AP).

Office (2006), the Navy sought to transition personnel from overmanned to undermanned occupations. The net result is that the Navy and Air Force did not always succeed in meeting overall reenlistment goals. However, in recent years, retention has been excellent. Table 2.2 also shows that AC enlisted retention exceeded goals in 2010 and 2011 for all services, with the exception of some categories of personnel for the Air Force.

Turning now to the RC, Table 2.3 shows enlisted recruiting results for the Selected Reserve in each component. The Army National Guard, Army Reserve, and Air National Guard did not meet their accession goals in 2005–2007, and the Navy Reserve did not meet its accession goal in 2005 and 2006. The Army National Guard did not meet its recruiting goal in 2010 and in 2011; however its overall strength increased from 358,391 in 2009 to 362,942 in 2010. Thus, its strength was several thousand greater than its authorized strength of 358,200 in 2010 (Hosek and Miller, 2011). Perhaps the Army National Guard decreased its recruiting effort in view of its high retention but did not decrease its reported recruiting goal. Nevertheless, across DoD, the RC have met or exceeded their overall enlisted accession targets since 2008.

Table 2.3 also shows the percentage of RC enlisted recruits who are HSDG and who score in AFQT Categories I to IIIA. DoD Instruction 1145.01 regarding recruit quality benchmarks applies to both the AC and RC. Across DoD, the RC have met the OSD benchmarks since 2006, although specific components did not always meet them between 2006 and 2008. However, since the economic downturn began, reserve recruit quality has been strong. Every component has exceeded the benchmarks, suggesting that the RC have been able to become more selective as recruit supply has improved.

Reserve attrition, defined as the percentage of enlisted reservists participating in a Selected Reserve unit in the previous year but not in the current year, decreased dramatically after 2008. Table 2.4 reports the degree to which the RC met or exceeded their attrition ceiling targets. For

Table 2.3
Selected Reserve Component Enlisted Recruiting

	2005	2006	2007	2008	2009	2010	2011
Percentage of Accessions Goal Achieved							
Army National Guard	80	99	95	103	100	95	95
Army Reserve	84	95	101	106	105	104	106
Navy Reserve	88	87	100	100	101	100	100
Marine Corps Reserve	102	100	110	100	122	125	100
Air National Guard	86	97	93	126	106	109	108
Air Force Reserve	113	106	104	105	109	105	103
DoD	N/A	97	98	105	104	101	100
Percentage in AFQT Categories I–IIIA							
Army National Guard		57	57	59	76	68	70
Army Reserve		59	57	58	63	71	69
Navy Reserve		73	66	62	75	79	82
Marine Corps Reserve		74	73	75	73	76	76
Air National Guard		77	75	75	77	78	98
Air Force Reserve		75	73	73	73	76	77
DoD		61	60	61	77	70	73
Percentage HSDG							
Army National Guard		91	91	91	94	95	92
Army Reserve		90	86	89	97	100	97
Navy Reserve		86	90	91	94	97	98
Marine Corps Reserve		96	96	97	98	100	100
Air National Guard		95	96	98	91	94	98
Air Force Reserve		96	99	99	99	100	100
DoD		91	91	91	95	96	95

SOURCE: OSD (MPP/AP).

example, in 2011, the Army National Guard had an attrition ceiling of 19.5 percent. It had an actual attrition rate of 18.8 percent, implying that actual attrition was 3.6 percent below the ceiling (or 4 percent rounded up, as shown in the table). Selected Reserve attrition in 2010 was 20 to 48 percent below the targeted attrition ceiling. In 2011, the attrition rate was between 4 and 31 percent below the ceilings set by the RC.

Table 2.4
Selected Reserve Component Attrition (percentage below attrition rate ceiling)

	2005	2006	2007	2008	2009	2010	2011
Army National Guard	–4	–11	–8	–13	–9	–20	–4
Army Reserve	–26	–32	–22	–32	–46	–48	–31
Navy Reserve	–20	–16	–22	–23	–40	–47	–23
Marine Corps Reserve	–33	–22	-22	-22	-26	–24	–17
Air National Guard	–23	–18	–20	–23	-30	-31	–17
Air Force Reserve	–26	–24	–12	–2	–18	–29	–10

SOURCE: OSD (MPP/AP).

CHAPTER THREE

Changes in the ECI and Basic Pay, 2000–2011

Overview

Military pay has risen faster than benchmarks:

- Basic pay is up 45 percent since 2000, substantially more than the ECI (up 33 percent) and the Consumer Price Index (CPI) (up 31 percent).
- Inflation-adjusted basic pay is far higher now than it was in 2000 (up 15 to 24 percent for enlisted personnel and up 7 to 20 percent for officers).
- The increases for RMC and total cash pay are even more dramatic. RMC is up about 40 percent for enlisted and 25 percent for officers.
- In contrast, real civilian pay has declined, dropping –4 percent to –8 percent in most groups.
- As a result, military pay is now well above the 70th percentile for civilian wages in most enlisted and officer age and education groups.
- Also, the cost of health care has increased rapidly in the civilian sector but remains at zero for service members and at quite low cost for their families (see Appendix B).
- Moreover, civilians have to contend with the risk of unemployment, which at present reduces the expected civilian wage by several more percentage points (see Appendix C).

Pay Changes Since 2000

In response to strains in recruiting and retention in the late 1990s, the National Defense Authorization Act of 2000 increased basic pay by 4.8 percent, restructured the basic pay table by providing higher pay increases for certain years of service and grades, and authorized increasing basic pay by half a percentage point above the ECI through FY 2006 (for example, if the ECI increase was 3 percent, the authorized basic pay increase would be 3.5 percent).[1] The 4.8 percent increase and the targeted increases led to an FY 2000 average pay increase of 6.2 percent (Goldich, 2005).

[1] There are several ECI series. The one used in adjusting basic pay is the 12-month percentage change in private industry wage and salary ECI for the third quarter of the calendar year, which is the end of the federal fiscal year. A lagged value of this ECI is used; for example, the ECI for the third quarter of 2009 was used in developing a basic pay adjustment for FY 2011.

The higher-than-ECI increases in basic pay actually continued to FY 2010. This helped to sustain recruiting and retention during wartime.[2] Frequent and long deployments to Iraq and Afghanistan stressed recruiting and retention, especially between 2005 and 2008 (Asch et al., 2010; Hosek and Martorell, 2009), but increases in basic pay, basic allowance for housing,[3] and enlistment and reenlistment bonuses enabled the services to meet their manning requirements. The increase in basic pay reverted to ECI in FY 2011. The current outlook is probably for basic pay increases equal to the percentage increase in ECI, the "default" in legislation (see Appendix A), although Congress may choose otherwise.

We next present tabulations of the percentage changes in ECI, basic pay, and the cost of living since 2000. We further consider the changes in RMC, total cash pay (defined below), and civilian pay for different age and education groups. We find that military pay has grown faster than ECI, the cost-of-living, and the wages and salaries of civilian workers comparable in age and education levels to those serving in the military.

Table 3.1 shows the percentage increase in basic pay since 2000. The table also includes the Consumer Price Index for urban consumers (CPI-U), with 2000 as the base year. Table 3.2 uses military pay files to show the actual increase in basic pay, including the restructuring adjustments to the basic pay table and the increase in RMC and total pay.

Table 3.1
ECI and Basic Pay Changes, 2000–2011

Year	ECI Annual Percentage Change	Basic Pay Annual Percentage Change	ECI Percentage Change from 2000	Basic Pay Percentage Change from 2000	CPI-U Percentage Change from 2000
2001	3.5	3.7	3.5	3.7	3.7
2002	3.1	4.6	6.7	8.5	4.9
2003	3.0	4.0	9.9	12.8	7.6
2004	2.6	3.6	12.8	16.9	9.7
2005	2.3	3.5	15.4	21.0	13.0
2006	3.0	3.1	18.8	24.7	17.5
2007	3.4	2.8	22.9	28.2	19.9
2008	2.9	3.5	26.4	32.7	25.0
2009	1.4	3.9	28.2	37.9	25.1
2010	1.6	3.4	30.2	42.6	28.4
2011	1.7	1.4	32.5	44.5	30.5

SOURCE: Bureau of Labor Statistics website. The basic pay increase for 2001 is from Goldich (2005).
NOTE: The basic pay increases shown are the statutory increases and do not account for additional increases resulting from restructuring of the pay table.

[2] A question is whether higher-than-ECI increases should have continued to FY 2010, given the recession and the higher-than-ECI increases throughout FY 2000–2009. However, from the outlook of FY 2009 when the FY 2010 increase was being discussed, policymakers might have expected the recession to abate; seen in retrospect, the recession deepened and lengthened.

[3] As explained on the OSD Military Compensation website, "The Secretary of Defense announced a major FY2001 Budget initiative to eliminate out-of-pocket costs for off-base housing in the United States. This action reduced service members' out-of-pocket costs for housing from an average of 18.8 percent of monthly housing costs in 2000 to 15 percent in 2001, with continued reductions each year thereafter. Average out-of-pocket costs were entirely eliminated in 2005."

Table 3.2
Percentage Increase in Median Real Basic Pay, RMC, and Total Cash Pay from 2000 to 2010, Enlisted Personnel and Officers, by Age Group and Service

	Age Group	Army	Navy	Air Force	Marine Corps
		Enlisted			
Basic pay	18–22	15	16	13	16
	23–27	13	14	13	18
	28–32	17	23	21	24
	33–37	15	20	20	20
RMC	18–22	33	37	28	24
	23–27	41	49	41	47
	28–32	40	46	42	54
	33–37	33	40	41	40
Total pay	18–22	46	40	22	29
	23–27	41	42	36	49
	28–32	40	41	40	47
	33–37	33	38	41	36
		Officers			
Basic pay	23–27	10	9	20	7
	28–32	11	11	13	12
	33–37	11	14	16	13
RMC	23–27	24	22	29	22
	28–32	24	24	23	26
	33–37	25	25	27	26
Total pay	23–27	27	22	29	22
	28–32	25	23	25	25
	33–37	25	26	26	23

SOURCE: Authors' tabulations based on military pay files and Bureau of Labor Statistics for CPI-U to adjust for cost-of-living increases.

The ECI increased by 33 percent from 2000 to 2011, and the CPI-U increased by 31 percent. Basic pay grew by 45 percent, or 12 percent faster than the ECI and 14 percent faster than the cost of living.

Table 3.2 reports real (inflation-adjusted) basic pay, RMC, and total pay from 2000 to 2010, by age group and service and for officers and enlisted personnel. *Total pay* is the sum of RMC and all other cash pays and allowances, such as special and incentive pays and cost-of-living allowances. The table uses the median values of military pay and the calculation of RMC factors in the service member's dependents' status. The table extends only to 2010 because military pay files were not available for 2011.

The main findings of Table 3.2 are that real basic pay increased over the decade for officers and enlisted personnel in every age group and service, and real RMC and total pay increased even more. Real RMC increased 24 to 54 percent for enlisted personnel and 22 to 29 percent for officers. The increase in total pay was about the same as the increase in RMC for officers but slightly lower for enlisted personnel. The faster growth of RMC than basic pay resulted from increases in the basic allowance for housing. Real basic pay increased 13 to 24 percent for enlisted personnel and 7 to 20 percent for officers. Basic pay changes for some age groups were higher than for other groups, which, as mentioned, reflects the restructuring of the basic pay table that targeted larger increases to pay cells corresponding to service members who had more experience or were promoted relatively fast. The increase by age group also could potentially be affected by changes in the average pay grade of an age group, but tabulations (not shown) indicate little year-to-year variation in average pay grade and no trend up or down.

While real military pay increased over the last decade, real civilian pay as measured by the CPS did not. Table 3.3 presents the percentage change in median real civilian wage from 2000 to 2009 by age and education. The civilian wages for 2009 are from the latest available March CPS, March 2010, for full-time, full-year male workers, defined as having 35 or more hours of work per week and 35 or more weeks of work per year. As seen, there is no case of an increase in the median civilian wage, and the decreases range down to 8 percent. But note that the ECI increased by 3 percent in real terms from 2000 to 2009 (see Table 3.1).[4] Chapter Four discusses why the ECI tends to overstate the changes in civilian wages during this period.

The 70th Percentile

The 9th Quadrennial Review of Military Compensation (QRMC) argued that RMC should reach at least the 70th percentile of the earnings of comparable civilians (Department of Defense, 2002). RMC has traditionally been considered the analog of civilian pay and includes basic pay, basic allowance for housing, basic allowance for subsistence, and the federal tax advantage of the allowances, which are tax free.[5] The 9th QRMC's position was based on the input of past commissions and study groups, as well as studies of military pay and recruiting and retention outcomes.

As discussed by the 9th QRMC, the 1948 Hook Commission report set the terms of the debate by establishing that military compensation rates should be based on comparisons between military and private-sector pay among those with similar levels of responsibility (Department of Defense, 1948). The President's Commission on an All-Volunteer Force in

Table 3.3
Percentage Change in Median Real Civilian Wage from 2000 to 2009, by Age and Education

Age Group	High School	Some College	Bachelor's Degree	Above Bachelor's Degree
18–22	–1	–6	–	–
23–27	–5	–3	–8	0
28–32	–4	–4	–4	–5
33–37	–2	–2	–4	–6

SOURCE: Authors' tabulations based on March CPS.

[4] The ECI increase from 2000 to 2009 in Table 3.1 is 28.2 percent, and the CPI-U increase is 25.1 percent. Hence, the inflation-adjusted ECI increased by 3 percent (28.2 – 25.1 = 3.1).

[5] The 10th QRMC recommended using military annual compensation, a metric that includes RMC as well as state and FICA tax advantage, the benefit of avoiding the cost of health care, and the value of the military retirement benefit. (Department of Defense, 2006).

1970 acknowledged that higher levels of military pay relative to pay for comparable civilians may be needed because of the hazards and other conditions of military service (Department of Defense, 1970). Thus, the ability to meet recruiting and retention targets in light of the nature of military service and the overall health of the all-volunteer force should be incorporated into analysis of the adequacy of military compensation.

Analysis of military and civilian pay in the 1990s argued that the selectivity and rigors of military service called for above-average pay and that pay around the 70th percentile had historically been necessary to enable the military to recruit and retain the required quality and quantity of personnel (see Asch, Hosek, and Warner, 2001). From the standpoint of recruiting, Sellman (2004) states that DoD's recruit quality benchmarks for the armed services of 60 percent in AFQT Categories I–IIIA and 90 percent HSDG were chosen as the minimum acceptable values based on the cohorts serving in 1990–1991. These cohorts produced satisfactory performance during Operations Desert Storm and Desert Shield. Their level of pay was at about the 70th percentile; junior enlisted and officer pays during the 1990s were at about the 70th percentile of comparable civilian pay between 1993 and 1999 (Hosek and Sharp, 2001). While the education benchmark is stated in terms of HSDG status, the 9th QRMC concluded that the appropriate comparison group for enlisted personnel was no longer just those with a high school diploma but also those with some college, since the military recruited from the college-bound youth market and a large fraction of the enlisted force has some college. Thus, the analysis of pay comparability should compare military pay for enlisted personnel to the 70th percentile of pay of comparable civilians with some college. Similarly, the comparison group for officers is civilians with a bachelor's degree or higher.

In sum, the origin of the 70th percentile as a focal point for military pay comparisons ties to the military-civilian pay levels for the cohorts serving in 1990–1991 at the time of Operations Desert Storm and Desert Shield. However, the 70th percentile is not a goal in and of itself. Compensation should be set high enough to attract and retain the quantity and quality of personnel the services require, and the level of compensation necessary to do this may or may not be at the 70th percentile.

Median RMC Compared with Civilian Wages

Analysis for the 11th QRMC finds that, from 2001 to 2009, RMC for enlisted personnel grew from the 77th percentile to the 85th percentile, and RMC for officers grew from the 76th to the 84th percentile (Grefer et al., 2011). The present report extends that analysis to compare RMC with civilian pay for different education and age groups, by branch of service.

Figures 3.1 and 3.2 are based on March Current Population Survey data and military pay files. Figure 3.1 shows civilian weekly wages by percentile from 2000 to 2009 for percentiles 30, 40, 50, 60, 70, 80, and 90 and overlays the median RMC. Because RMC varies somewhat by service (largely due to cross-service variations in promotion speed), we show a separate chart for each service. Wages and RMC are stated in 2010 dollars. Panels A and B, which refer to enlisted personnel, are for civilian workers ages 23–27 with a high school diploma and for those with some college. Panels C and D, relevant to officers, are for civilian workers ages 28–32 with a bachelor's degree and for those with more than a bachelor's degree.

In 2000, RMC was around the 60th percentile of civilian weekly wages for high school graduates and the 50th percentile for those with some college. In 2009, RMC was between

Figure 3.1
Real Civilian Wages and Median RMC by Service, 2000–2009, in 2010 Dollars

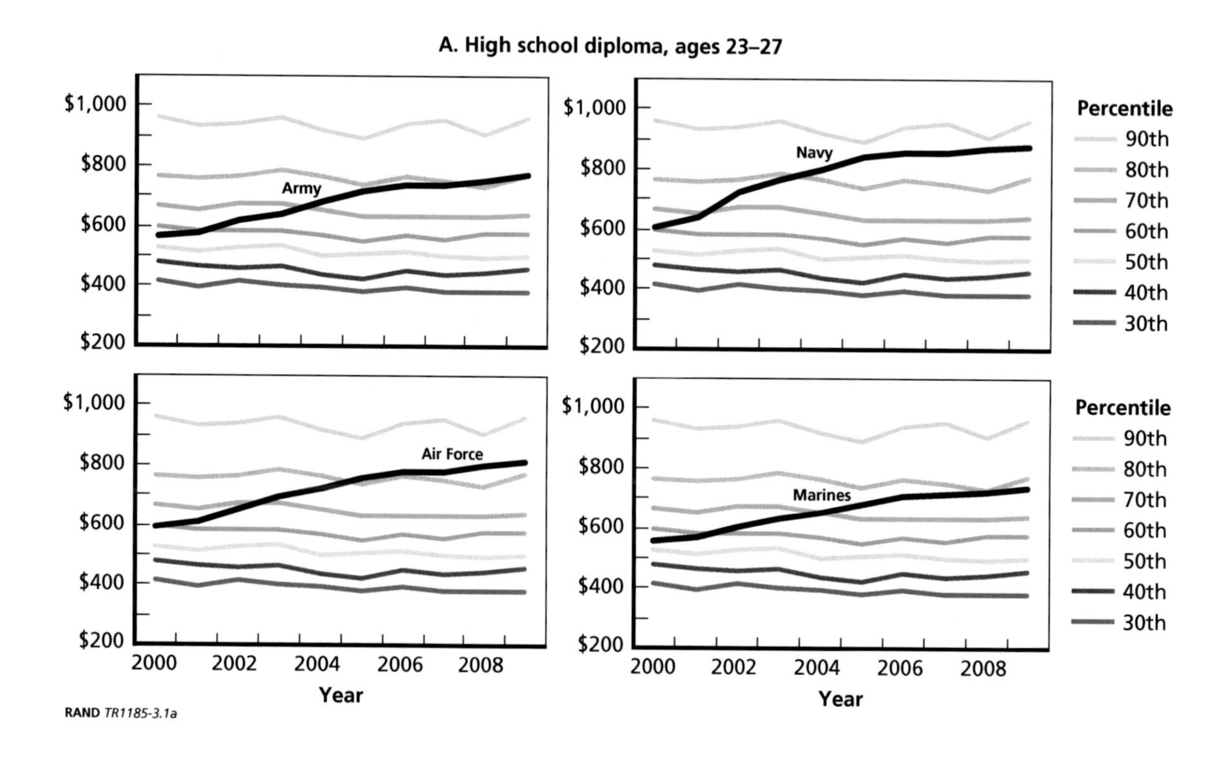

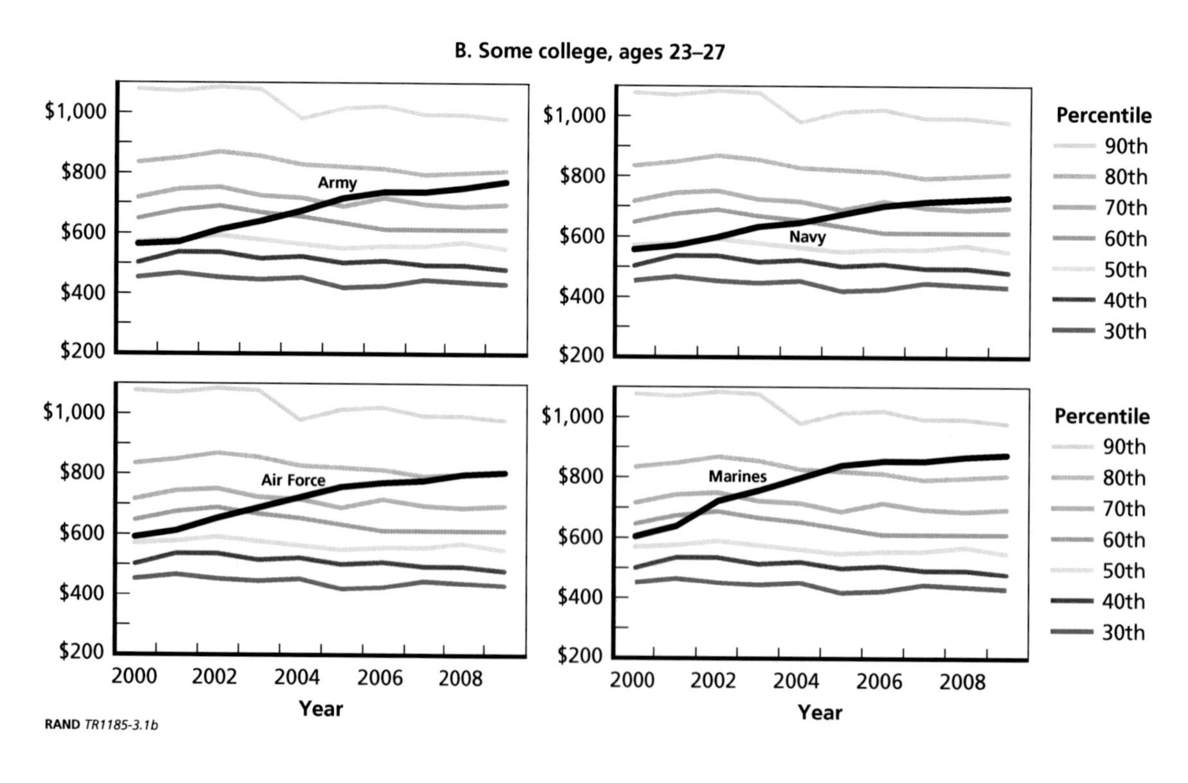

Figure 3.1—Continued

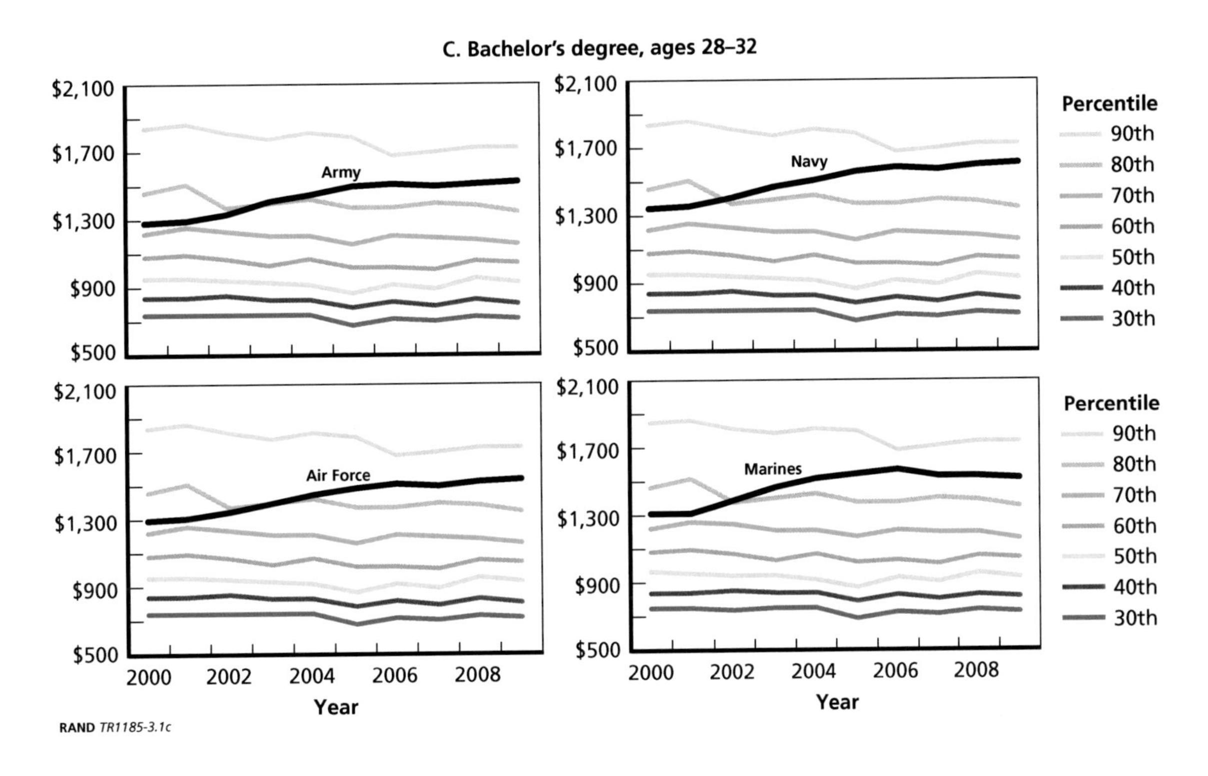

RAND *TR1185-3.1c*

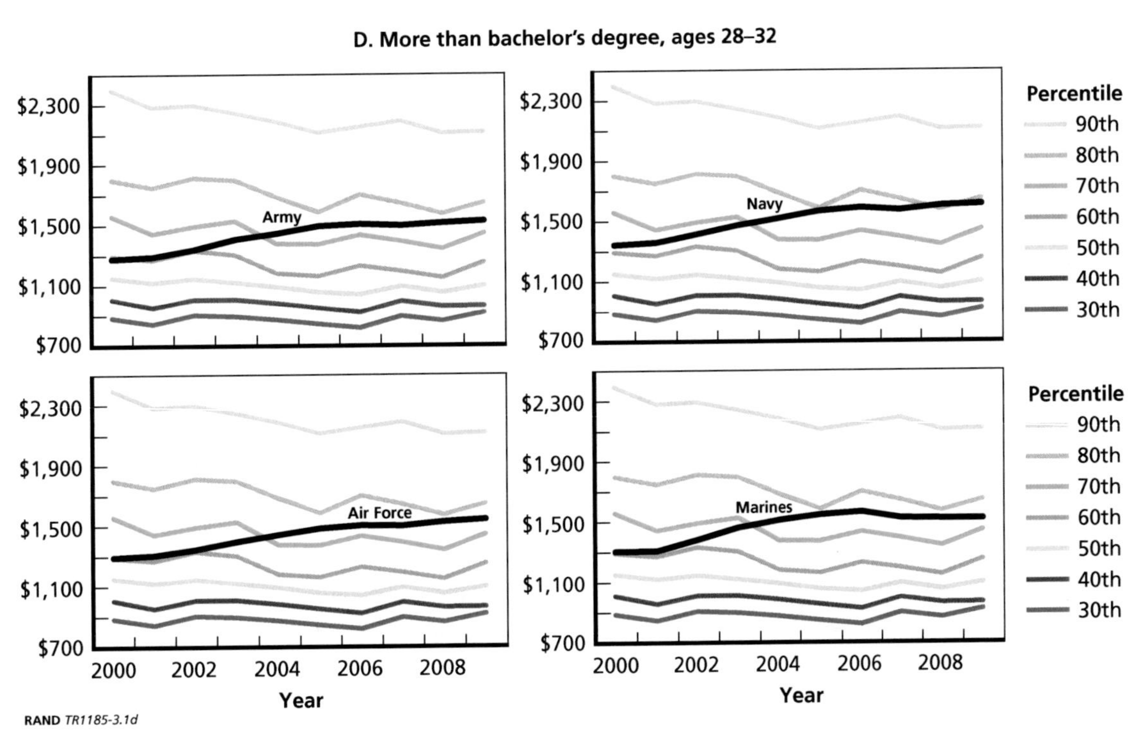

RAND *TR1185-3.1d*

Figure 3.2
Real Civilian Wages and Median RMC for Those with More Than a Bachelor's Degree, by Service, 2000–2009, in 2010 Dollars

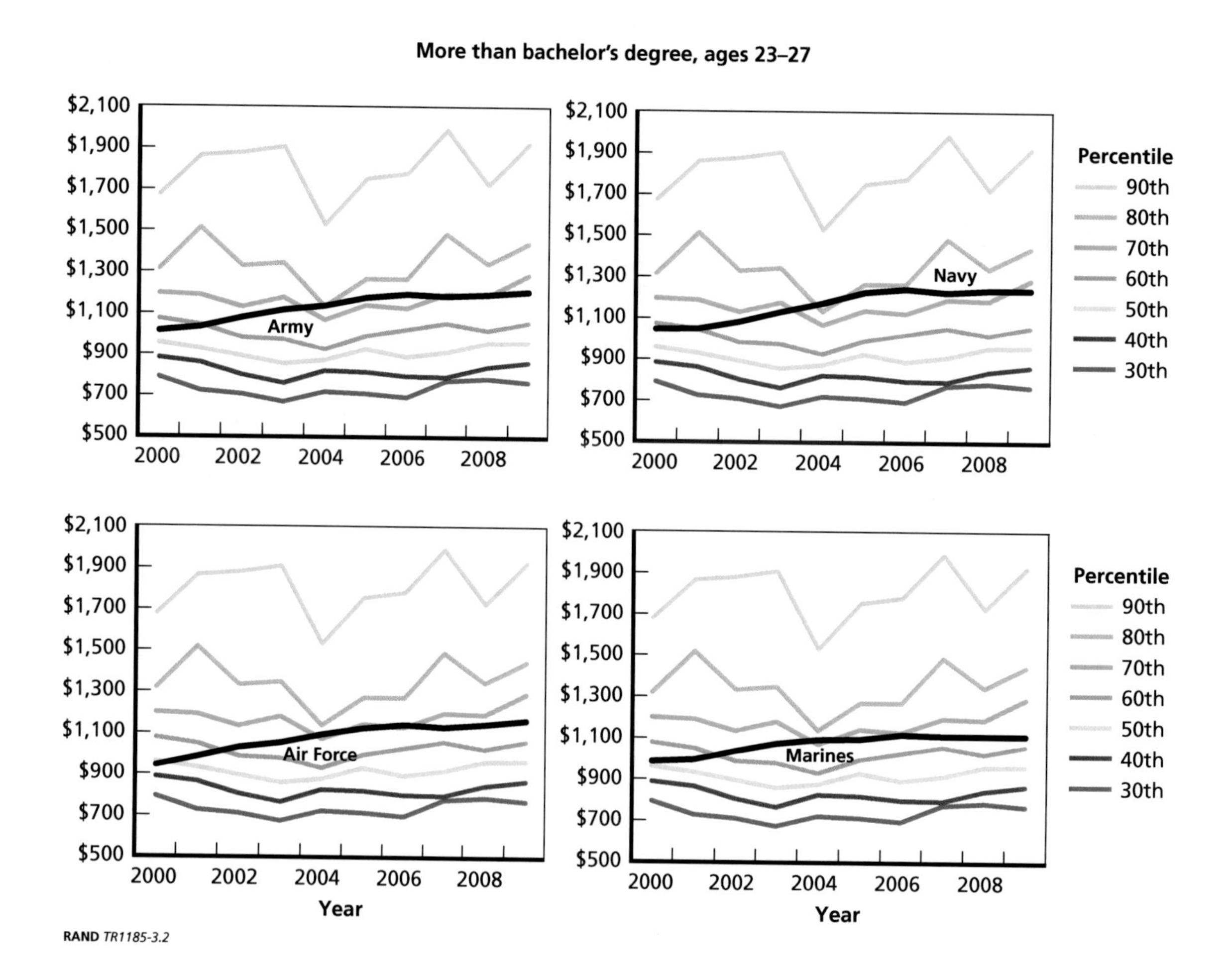

RAND *TR1185-3.2*

the 80th and 90th percentile for the Army, Navy, and Air Force and just below the 80th for the Marine Corps for high school graduates, and it was near the 80th percentile for those with some college.

Officer RMC was at the 70th wage percentile in 2000 for civilians with a bachelor's degree, and it was at the 85th percentile in 2009. For civilians with more than a bachelor's degree, officer RMC rose from around the 60th percentile in 2000 to the 75th percentile in 2009.

These results are consistent with Grefer et al.'s (2011) finding that RMC is now well above the 70th percentile for enlisted and officer personnel. Figure 3.1 also makes clear that real civilian wages had small decreases across all percentiles from 2000 to 2009, though real wage decreases were somewhat greater at the highest percentiles for workers with a bachelor's degree or more than a bachelor's degree. Comparisons were also made for age groups 18–22, 23–27, 28–32, and 33–37; the results were similar and are not shown.

However, the results differ for workers ages 23–27 with more than a bachelor's degree. These people are "fast-burners" who complete a master's degree or higher right after their bachelor's degree. This group might be especially relevant to judging the adequacy of military pay with respect to the type of person the services want to enter the officer corps. Figure 3.2 compares RMC for this group with civilian wage percentiles. As before, civilian wages are for full-time, full-year male workers, defined as having 35 or more hours of work per week and 35 or more weeks of work per year. RMC by service was between the 50th and 60th percentiles in 2000 and at or above the 70th percentile in the middle of the decade, but was below the 70th percentile in 2009. This finding may merit further attention, depending on whether the services find the supply and quality of junior officers to be adequate.

CHAPTER FOUR

A Comparison of ECI and Median Weekly Wage Increases Since 2000

In the previous chapter, we found that the real ECI increased but real median civilian wages decreased from 2000 to 2009 for the age and education groups we examined. This chapter argues that these trends differ because of the index nature of the ECI. Civilian wage trends that are not based on an index are likely to be a more accurate indicator of the civilian opportunity wage of military recruits and personnel.

The ECI is designed to track changes in the cost of labor while holding fixed the number of workers in each narrowly defined occupational category.[1] More specifically, the ECI is a Laspeyres index of employment costs. In a Laspeyres index, a base period is selected, and quantity weights are defined for that period. Here, the base period is 2002, and the quantity weights are determined by the number of workers employed in each occupational category or "cell." The ECI denominator is the sum, by cell, of base period employment multiplied by base period wages; the numerator is the sum, by cell, of base period employment multiplied by current period wages. As wages change from year to year, the ECI tracks the change in the cost of the particular bundle of labor defined by the 2002 employment weights.

The change in the ECI can differ from the change in the median (or average) wage if the mix of employment changes from the mix at the base period. Actual employment might increase in some cells and decrease in others. For instance, an establishment's product mix may change, some employers may go out of business, and new employers may appear. These changes in employment will typically be a combination of secular trends and cyclical changes. Employment in service industries has increased for many decades, and employment in manufacturing has decreased. The decade of the 2000s began in a mild recession followed by a gradual expansion, but the worst recession on record began in 2008 and has continued since then. Related to

---

[1] ECI data come from the National Compensation Survey (NCS), a survey of establishments. According to the *Handbook of Methods* (HOM), "NCS data are collected from probability samples selected in three stages: (1) a probability sample of geographic areas, (2) a probability sample of establishments within sampled areas, and (3) a probability sample of occupations within sampled establishments." (HOM, p. 2) The HOM states that the Bureau of Labor Statistics (BLS) asks about the following in initial and subsequent visits to establishments: What is the primary business activity of the establishment; what types of occupations does the establishment employ; how many employees are there in each sampled job that is matched to an occupational description; do workers in the matched, sampled occupation work full- or part-time, are they union or nonunion, and are they paid by time or incentive; what are the employees in the sampled, matched occupation paid; what are the duties and responsibilities of the job; how many hours does the employee work; and what type of benefits do the employees receive? (HOM, p. 4.) In determining the work level in the job, the NCS considers four factors: knowledge, job controls and complexity, contacts, and physical environment. "Each factor consists of several degrees, each with an associated description and number of points. Generally, the greater the consequence, complexity, or difficulty of the factors, the higher is the number of points assigned." (HOM, pp. 6–7.) Base period employment is 2002. (HOM, p. 12.)

the change in employment, workers may move from one job to another, e.g., from a low-paying job to a high-paying job or vice versa. The ECI is constructed to hold employment constant with base-period weights, so changes in the mix of employment are not captured by the ECI.

To give a cyclical example of how the ECI change might differ from the median wage change, suppose establishments shed workers in an economic downturn and that workers who find new jobs typically receive a lower wage than they did before. In the framework of the ECI, we can envision these workers moving to a lower wage cell, and this would tend to decrease the median wage. But the wage paid in any cell might be little affected as workers come and go, at least in the short run. In fact, wages might show some increase from year to year even though workers who change jobs and take lower-wage jobs might cause the median wage to decrease. Again, the ECI is designed to be an index of employment cost for a given bundle of labor; it is not designed to track the median (or average) wage among workers in the labor force.

By comparison, the CPS weekly wage comes from respondents' reports of their usual weekly wage, and the published weekly wage series is based on the median weekly wage of the usual weekly wages reported by full-time workers. The CPS weekly wage is not an index and is therefore not affected by index issues such as base period employment. The CPS will reflect trends in the median wage, including the influence of movements to higher or lower wage jobs.

Tables 3.1 and 3.3 showed that from 2000 to 2009 the inflation-adjusted ECI increased by 3 percent,[2] while real civilian wages decreased by up to 8 percent for specific age and education groups. The civilian wage data were from the March CPS and the available coverage ended with 2009 (from the March 2010 CPS). The median weekly wage series, however, is currently available to the second quarter of 2011, as is the ECI. Table 4.1 presents a comparison of the percentage change since 2000 of the inflation-adjusted ECI (as before, the ECI is for wages and salaries of private industry workers) and the inflation-adjusted median weekly wage, by education. The published median civilian wage is by education only, not by age group and education.

Table 4.1
Percentage Change in Real ECI and Real Median Civilian Wage by Education, 2000–2011

Year	ECI	Less Than High School	High School	Some College	Bachelor's	More Than Bachelor's
2001	0.1	1.8	–0.8	0.3	0.9	–1.7
2002	2.0	1.2	0.4	2.4	2.3	1.1
2003	2.6	–0.3	0.8	0.5	3.9	1.6
2004	2.9	–0.1	4.0	2.3	1.4	–0.4
2005	2.2	–1.2	2.3	–1.5	0.3	–0.7
2006	1.4	–3.3	–0.6	–3.3	1.9	–0.9
2007	2.4	–0.9	–2.5	–1.5	4.6	–3.6
2008	0.9	–3.5	–3.0	–2.7	–2.4	–1.2
2009	3.4	1.6	–0.3	–3.3	2.3	2.8
2010	2.9	–7.8	–3.1	–2.9	–0.1	1.1
2011	1.5	–6.5	–3.8	–5.4	–2.3	–2.2

SOURCE: Bureau of Labor Statistics for median civilian wage.

NOTE: Cell entries are adjusted by CPI-U for the 12-month change ending June 30.

[2] In Table 3.1, we saw that from 2000 to 2009 the ECI increased by 28.2 percent and the CPI-U increased by 25.1 percent, implying that the ECI rose 3 percent more than the cost of living.

Table 4.1 is consistent with the earlier finding: The real ECI increased from 2000 to 2011 and the real median wage decreased. The real ECI increased only slightly during the decade relative to 2000, and by 2011 the real ECI had grown by only 1.5 percent. Still, the real median civilian wage decreased at all levels of education. The decrease ranged from –6.5 percent to –2.2 percent, and the decrease was larger for those at lower levels of education (less than high school and high school) than for those with some college or at least a college degree. Further, the decrease has grown more rapidly since the onset of the recession in 2008. This comparison between the ECI and the median wage suggests that the ECI mutes the full extent of change in the civilian wage. In current circumstances, the ECI shows a small real increase in the face of real decreases, yet during an upswing the opposite could occur, i.e., the ECI might show smaller wage growth than CPS wage data reveal.

CHAPTER FIVE

Conclusion

An Opportune Time for Action

Whether the rate of increase in military pay should be slowed is a policy decision. In fact, as mentioned at the beginning of this report, DoD plans to limit military pay raises beginning in 2015. The data we have presented show that recruiting quality and retention rates are extraordinarily high and that military pay has grown relative to civilian pay. Military pay now stands well above the 70th percentile on the civilian wage distribution. In addition, the nation is planning to reduce military personnel strength and seeking to trim the federal budget. Taking these factors into consideration, we think slower growth of military pay relative to the ECI is advisable. It is unlikely to hurt capability and readiness; where needed, bonuses and special pays can be used to manage recruiting and retention and do so more cost-effectively than across-the-board pay actions.

The AC and RC have made their recruiting and retention targets since 2009. This success is likely a result of past increases in military pay and the current recession. Both basic pay and RMC increased relative to civilian wages throughout the decade. Adjusting for inflation, civilian wages decreased by several percent since 2000, basic pay increased by 13 percent, and RMC increased 24 to 54 percent for enlisted personnel and 22 to 26 percent for officers. In addition, the unemployment rate is now high (around 8.3 percent) and has increased substantially at all levels of education; further, long spells of unemployment are common—indicating a lack of job openings despite active job search. Many workers work part-time for economic reasons, meaning slack work, poor economic conditions, and the unavailability of full-time work.

Some Options for Slowing the Increase in Basic Pay

If the rate of increase in military pay were to be decreased, how should it be done? We consider three options: (1) a one-time increase in basic pay set at half a percentage point below the ECI, (2) a one-year freeze in basic pay (no change), and (3) a series of below-ECI increases, such as ECI minus half a percentage point for four years.

- **A one-time increase in basic pay of half a percentage point below ECI.** The Congressional Budget Office (CBO) has estimated that a one-time increase in basic pay of half a percentage point *above* ECI, if implemented in 2011, would increase costs by about $360 million in the first year, $2.4 billion in five years, and $5.2 billion in ten years (CBO,

2010). One might assume that the same increase *below* ECI would provide cost savings of about the same amount.

- **A one-time pay freeze.** A pay freeze in FY 2013 would mean that basic pay would not change although the (lagged, FY 2011) value of the ECI was 1.7 percent (Table 3.1). A pay freeze of this magnitude is less than the decrease in the expected civilian wage (adjusted for unemployment) that has already occurred since 2008 (see Appendix C), namely, 2.5 percent for workers with a bachelor's degree or more and 6 percent for those with a high school diploma or some college. As a rough estimate using the CBO's numbers, a basic pay freeze might provide cost savings 3.4 (= 1.7/0.5) times more than the one-time 0.5-percent decrease, or $1.2 billion in 2013, $8.2 billion in 2013–2017, and $17.7 billion in 2013–2022.
- **Four years of basic pay increases set at ECI minus half a percentage point.** A repeated decrease in basic pay of half a percentage point below ECI over four years (2012–2015) would create cost savings of $280 million the first year, $5.8 billion over 2012–2016, and $17.5 billion over 2012–2021 (CBO, 2011). Thus, to a first approximation, the savings from the pay freeze and the four-year program of ECI minus half a percentage point are about equal.

To put this in the context of overall defense spending, DoD outlays in 2010 were $689 billion, of which $151 billion (22 percent), were for military personnel, and the CBO projects DoD spending to decrease from about 4.5 percent of gross domestic product in 2010 to 3.8 percent in 2021, or by about one-sixth (CBO, 2011). The CBO further projects, as an exercise, that reducing the growth in DoD outlays by one percentage point annually would generate $286 billion in savings over 2012–2021 (CBO, 2011). The $17.5 billion in savings from slower basic pay growth or a one-time pay freeze amounts to 6 percent of these projected savings while a one-time half a percentage point decrease below ECI would be almost 2 percent of these projected savings. Thus, the potential for savings is less with the first option than with the second and third options.

A decision to slow the increase in military pay, however implemented, will carry political costs. Service members and their families are likely to find a decrease in the rate of growth of military pay unwelcome. A reduction during wartime might be interpreted as a signal of waning support for the war effort and a lack of appreciation for the personal sacrifices of service members and their families. Morale, retention, and recruitment might suffer as a consequence.[1]

The political costs associated with a pay freeze (option 2) and a four-year schedule of below-ECI pay increases (option 3) are likely to be greater than those of a one-time below-ECI pay increase (option 1) simply because the overall reduction in pay is larger for options 2 and 3 than for option 1. In fact, the literature on loss aversion (Kahneman and Tversky, 1979) suggests that the magnitude of an individual's reaction to a loss increases disproportionately with the magnitude of the loss itself. As for the relative political costs carried by options 2 and 3, it is not clear which of the two would generate the more negative effect. On the one hand, the

[1] Slowing growth in military pay by a large amount or over an extended period may adversely affect retention or recruiting in certain occupational specialties. The CBO has remarked on the need to have bonus funds available to counteract such outcomes and added that legislative changes might be needed that relax restrictions on the maximum size of bonus awards. More generally, "To alleviate any effect on retention during those four years [2012–2015], the service branches could increase bonuses for enlistment and reenlistment, step up recruiting efforts, or offer other benefits to service members." (CBO, 2011, p. 76.)

reaction to a pay freeze may be stronger than the reaction to a four-year schedule of smaller payments simply because a pay freeze represents a large and sudden change relative to service member expectations. On the other hand, the four-year program (option 3) may cause service members to feel uncertain about their income stream because policymakers could elect to change the program after the initial year. Service members' responses to uncertainty can vary widely, but for those who are more risk-averse, the uncertainty may magnify any adverse reaction.

The three options also vary with respect to their political feasibility and their potential for hedging against uncertainty. Options 1 and 2 may be more politically feasible because both involve a one-time change in military pay growth at a time when conditions favor such a policy decision. In particular, the high rate of unemployment in the civilian labor market, the robust state of military recruitment and retention, and planned drawdowns in personnel strength provide justification for a reduction in the growth of military pay. Option 3 may be more difficult to implement because any change in these conditions could cause the political will to implement the four-year program to falter. That being said, options 1 and 3 provide greater opportunity for hedging against uncertainty. Both of these options prescribe a smaller reduction in the growth of military pay in the short term and, as such, allow for adjustments to changing conditions in subsequent years. For instance, if retention and recruitment were to prove challenging in three years' time, and a reversal of slower pay growth were desired, option 3 could more easily accommodate the reversal because the full decline in growth (of the magnitude prescribed by option 2) would not yet have occurred.

The three implementation options presented invoke two trade-offs. First, there is the trade-off between cost savings and political costs. Second, there is the trade-off between political feasibility and hedging against uncertainty. How do the three options compare with respect to these factors? Option 1 (a one-time below-ECI pay increase) is more feasible politically and provides greater opportunity for adjusting to changing conditions, but it provides much less in the way of cost savings. Option 2 (a pay freeze) provides greater cost savings and does so quickly, but it may generate a larger adverse political reaction and provides less opportunity for hedging against uncertainty. Option 3 (four years of a below-ECI pay increase) also provides greater cost savings and, by implementing the savings over an extended period, provides greater opportunity for adjusting to changing conditions. However, option 3 may be more difficult to carry out and would likely carry large political costs because of the need to gain congressional passage repeatedly.

Since we are unable to measure and assess the relative importance of these factors rigorously, we do not recommend a specific implementation approach. Policymakers need to weigh the various factors in deciding on a course of action. As mentioned, DoD plans to limit basic pay increases starting in FY 2015, yet it may revise its position in the future, and Congress and the President might have still other approaches in mind. While we cannot recommend a specific means of implementation, we can recommend that pay growth be slowed relative to the ECI. Our analysis provides strong evidence for slowing growth given the stated goals of DoD and the services for quantity and quality and the current economic climate.

Interaction with Retirement Reform

Service members are aware of the policy discussion over military retirement benefits. The impact of a pay freeze or slowed pay growth may interact with changes in military retirement benefits. In thinking about policy changes, one should consider how retirement reform coupled with slowed pay growth would affect retention and recruiting. RAND has work under way on this question. Preliminary results were presented to OSD Compensation in the fall of 2011, and extensions of the model to consider the transitional effects on force structure are in progress, with results expected by September 2012.

Replace the ECI with the CPS Weekly Wage?

Our discussion also raises the question of whether adjustments to basic pay should be based on the CPS median wage rather than the ECI. The ECI is an index that describes the employment cost of a specific set of labor categories, whereas the CPS median wage tracks wage changes over time. Because the CPS median wage is not constrained by particular labor categories, it is more sensitive to changes in the civilian wage. We have presented data showing that in recent years the ECI has overestimated civilian wage growth and have argued that in a time of economic expansion the ECI might underestimate civilian wage growth. Since the CPS median wage is available on as timely a basis as the ECI, its use would not delay the development of defense personnel budget submissions. Moreover, models of individual enlistment and retention behavior include variables for the civilian opportunity wage; because the CPS median wage reflects wages at large, it is a better measure of the civilian opportunity wage than the ECI is. We recommend that the CPS median wage be used instead of the ECI as the basis for adjusting basic pay.

APPENDIX A

U.S. Code Title 37, Chapter 19, Section 1009, "Adjustments of Monthly Basic Pay"

(a) **Requirement for Annual Adjustment.**— Effective on January 1 of each year, the rates of basic pay for members of the uniformed services under section 203 (a) of this title shall be increased under this section.

(b) **Effectiveness of Adjustment.**— An adjustment under this section shall have the force and effect of law.

(c) **Equal Percentage Increase for All Members.**—

(1) An adjustment made under this section in a year shall provide all eligible members with an increase in the monthly basic pay that is the percentage (rounded to the nearest one-tenth of one percent) by which the ECI for the base quarter of the year before the preceding year exceeds the ECI for the base quarter of the second year before the preceding calendar year (if at all).

(2) Notwithstanding paragraph (1), but subject to subsection (d), the percentage of the adjustment taking effect under this section during each of fiscal years 2004, 2005, and 2006, shall be one-half of one percentage point higher than the percentage that would otherwise be applicable under such paragraph.

(3) In this subsection:

(A) The term "ECI" means the Employment Cost Index (wages and salaries, private industry workers) published quarterly by the Bureau of Labor Statistics.

(B) The term "base quarter" for any year is the three-month period ending on September 30 of such year.

(d) **Protection of Member's Total Compensation While Performing Certain Duty.**—

(1) The total daily equivalent amount of the elements of compensation described in paragraph (3), together with other pay and allowances under this title, to be paid to a member of the uniformed services who is temporarily assigned to duty away from the member's permanent duty station or to duty under field conditions at the member's permanent duty station shall not be less, for any day during the assignment period, than the total amount, for the day immediately preceding the date of the assignment, of the elements of compensation and other pay and allowances of the member.

(2) Paragraph (1) shall not apply with respect to an element of compensation or other pay or allowance of a member during an assignment described in such paragraph to the extent

that the element of compensation or other pay or allowance is reduced or terminated due to circumstances unrelated to the assignment.

(3) The elements of compensation referred to in this subsection mean—

(A) the monthly basic pay authorized members of the uniformed services by section 203 (a) of this title;

(B) the basic allowance for subsistence authorized members of the uniformed services by section 402 of this title; and

(C) the basic allowance for housing authorized members of the uniformed services by section 403 of this title.

(e) **Presidential Determination of Need for Alternative Pay Adjustment.—**

(1) If, because of national emergency or serious economic conditions affecting the general welfare, the President considers the pay adjustment which would otherwise be required by this section in any year to be inappropriate, the President shall prepare and transmit to Congress before September 1 of the preceding year a plan for such alternative pay adjustments as the President considers appropriate, together with the reasons therefor.

(2) In evaluating an economic condition affecting the general welfare under this subsection, the President shall consider pertinent economic measures including the Indexes of Leading Economic Indicators, the Gross Domestic Product, the unemployment rate, the budget deficit, the Consumer Price Index, the Producer Price Index, the Employment Cost Index, and the Implicit Price Deflator for Personal Consumption Expenditures.

(3) The President shall include in the plan submitted to Congress under paragraph (1) an assessment of the impact that the alternative pay adjustments proposed in the plan would have on the Government's ability to recruit and retain well-qualified persons for the uniformed services.

APPENDIX B

Health Care Cost Avoidance

Hosek et al. (2005) and Grefer et al. (2011) note that the military health benefit has become an increasingly relevant part of military compensation. Incorporating health cost avoidance into the pay comparison pushes up the military pay line by several percentiles (Hosek et al., 2005; Grefer et al., 2011). The services provide health care at no cost to service members and at extremely low cost to service members' dependents. In contrast, the cost of providing health care in the civilian economy has been increasing rapidly relative to other sectors of economic activity. The Kaiser Family Foundation survey of employer health benefits found that the annual average worker contribution to premiums in 2010 was $899 for single coverage and $3,997 for family coverage. (Kaiser, 2010). For civilians without employer-provided health care benefits, the cost of health care would be higher because employers often pay a sizable share of the premiums. Grefer et al. (2011) take this into account in developing their estimates.

Also, health care costs have risen rapidly since 2000, so the value of cost avoidance has also increased since then. Average annual worker contributions to premiums for single and family coverage were $334 and $1,619, respectively, in 2000. The approximately 150 percent increase in average annual premiums from 2000 to 2010 is obviously much greater than the 31 percent increase in the cost of living. The actual increase in health care costs is even greater, as individuals and families must make co-payments when receiving treatment.

APPENDIX C

Civilian Employment Conditions

Our tabulations of civilian pay are for full-time workers, but the recent recession has meant that workers are less likely to work full-time. Consequently, the expected full-time pay of civilians, defined as the full-time wage times the likelihood of full-time employment, has changed as well. This appendix describes changes in civilian employment and shows trends in expected civilian pay. However, because no series of the likelihood of full-time employment is available, we approximate the expected full-time wage series by using the trend in likelihood of either full- or part-time employment, which *is* available.

Since 2008, the U.S. economy has suffered the worst economic downturn since the great depression. Unemployment rates are high, and this is so at all education levels. The number of weeks people spend unemployed has increased, and many workers are working part-time for economic reasons.[1] Figure C.1 shows the movement in these series from 2000 to 2011.

The higher unemployment rate translates into a lower probability of either full- or part-time employment. This follows from the fact that among labor force participants, the fraction of employed workers equals 1 minus the fraction of unemployed workers. The year 2000 was a year of low unemployment, as were 2005, 2006, and 2007, and any of these years could be used as a base year for an index of the chance of employment. Taking 2000 as the base year where the probability of employment is indexed to 1, Figure C.2 shows the decrease in the indexed probability of employment since the recession began. As seen, the decrease occurred rapidly in 2008 and has remained at the new, lower level since then. The decrease was 2–3 percent for college-educated workers, 6 percent for both high school diploma graduates and those with some college, and 8 percent for those with less than a high school diploma.

[1] CPS technical documentation describes part-time work for economic reasons as follows: "Sometimes referred to as involuntary part-time, this category refers to individuals who gave an economic reason for working 1 to 34 hours during the reference week. Economic reasons include slack work or unfavorable business conditions, inability to find full-time work, and seasonal declines in demand. Those who usually work part-time also must indicate that they want and are available to work full-time to be classified as being part-time for economic reasons." (CPS, 2006, p. 5-3.)

Figure C.1
Unemployment Rate, Weeks Unemployed, and Number Working Part-Time for Economic Reasons

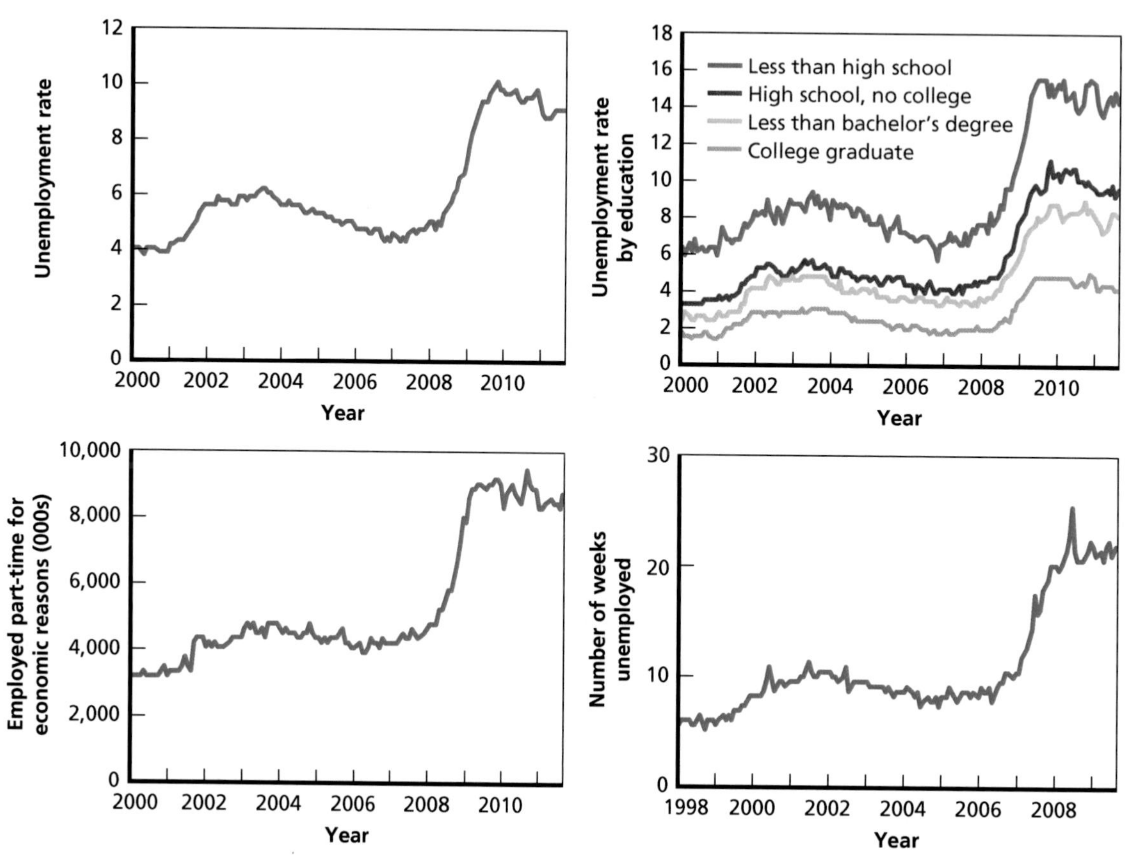

SOURCE: Bureau of Labor Statistics.
RAND *TR1185-C.1*

The CPS provides data on the median weekly wage of full-time workers,[2] and this wage can be multiplied by the indexed probability of employment to obtain an adjusted wage, which is a crude and probably conservative estimate of the expected weekly wage. That is, the expected weekly wage is probably lower. The estimate is crude because the weekly wage is for

---

[2] The CPS bases its weekly wage on data collected on earnings, as described in its technical report:

> Information on what people earn at their main job is collected only for those who are receiving their fourth or eighth monthly interviews. This means that earnings questions are asked of only one-fourth of the survey respondents each month. Respondents are asked to report their usual earnings before taxes and other deductions and to include any overtime pay, commissions, or tips usually received. The term "usual" means as perceived by the respondent. If the respondent asks for a definition of usual, interviewers are instructed to define the term as more than half the weeks worked during the past 4 or 5 months. Respondents may report earnings in the time period they prefer—for example, hourly, weekly, biweekly, monthly, or annually. (Allowing respondents to report in a periodicity with which they were most comfortable was a feature added in the 1994 redesign.) Based on additional information collected during the interview, earnings reported on a basis other than weekly are converted to a weekly amount in later processing. Data are collected for wage and salary workers, and for self-employed people whose businesses are incorporated; earnings data are not collected for self-employed people whose businesses are unincorporated. (Earnings data are not edited and are not released to the public for the 'self-employed incorporated.')

Figure C.2
Indexed Probability of Employment (2000 = 1.00)

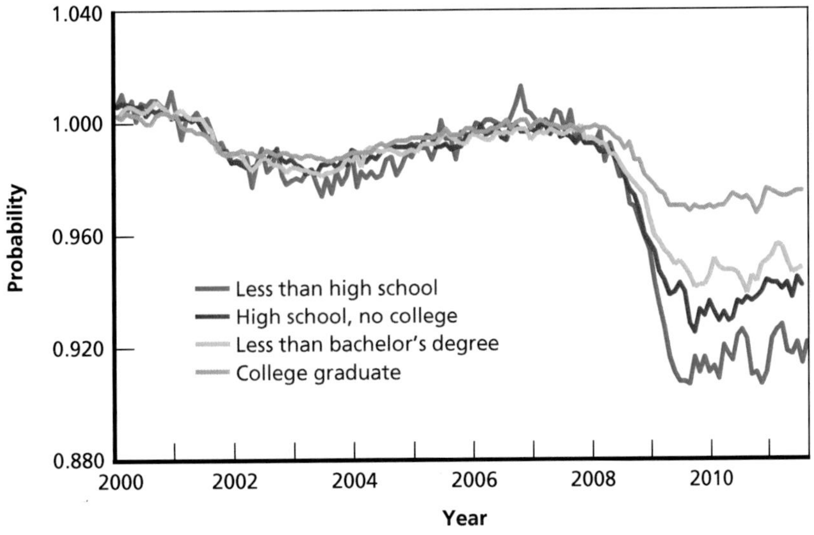

SOURCE: Bureau of Labor Statistics.
RAND *TR1185-C.2*

full-time workers, while the indexed probability of employment is for either full- or part-time employment. Given that the number of workers working part-time for economic reasons has greatly increased—climbing from about 4.5 million in 2007 to 8.5 to 9 million in 2009, 2010, and 2011—the probability of full-time employment has likely decreased faster than the probability of either full- or part-time employment. This implies that the adjusted wage shown here is likely to be a conservative estimate.

Figure C.3 shows two series, the median weekly wage and the median weekly wage multiplied by the indexed probability of employment, labeled as "adjusted." The two wage series, both adjusted for inflation (2010 dollars), did not differ much until the recession. Since then, however, the adjusted wage has fallen by $40–$50 per week for workers with less than a college education and $25 per week for college graduates. Thus, the expected pay of civilians has declined over the last decade, in small part because of declines in the real (inflation-adjusted) civilian pay of full-time workers (as shown in Table 3.3) but in large part because of the dramatic increase in the likelihood of being unemployed that has occurred since 2008.

These earnings data are used to construct estimates of the distribution of usual weekly earnings and median earnings. Individuals who do not report their earnings on an hourly basis are asked if they are, in fact, paid at an hourly rate and if so, what the hourly rate is. The earnings of those who reported hourly and those who are paid at an hourly rate is used to analyze the characteristics of hourly workers, for example, those who are paid the minimum wage. (CPS, 2006, p. 5-4.)

Figure C.3
Median Weekly Wage and Adjusted Median Weekly Wage, 2000–2011

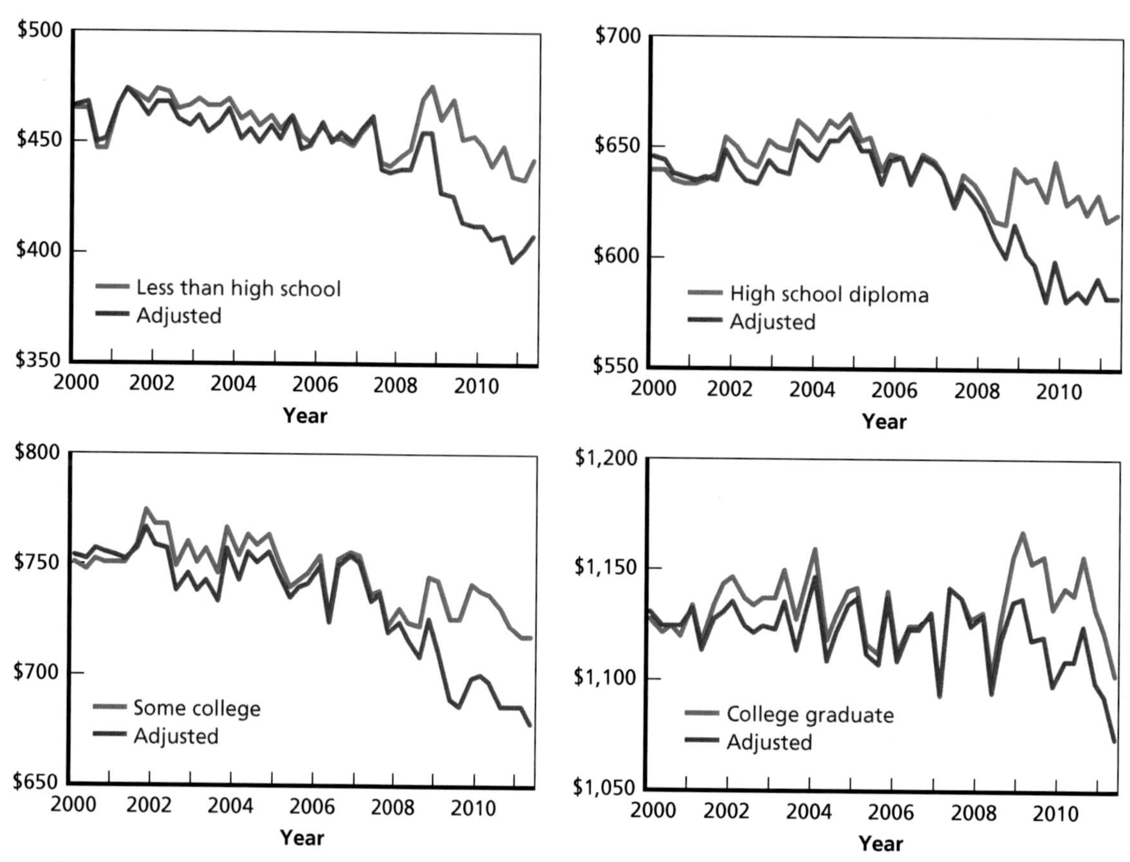

SOURCE: Bureau of Labor Statistics and authors' adjustment.
RAND *TR1185-C.3*

References

Asch, Beth J., Paul Heaton, James Hosek, Paco Martorell, Curtis Simon, and John T. Warner, *Cash Incentives and Military Enlistment, Attrition, and Reenlistment*, Santa Monica, Calif.: RAND Corporation, MG-950-OSD, 2010. As of April 26, 2012:
http://www.rand.org/pubs/monographs/MG950.html

Bureau of Labor Statistics, website, no date. As of April 25, 2012:
http://bls.gov/

———, *Handbook of Methods,* Chapter 8, "National Compensation Measures," 2011. As of April 26, 2012:
http://www.bls.gov/opub/hom/homch8.htm

CBO—*See* Congressional Budget Office.

Congressional Budget Office, *Recruiting, Retention, and Future Levels of Military Personnel*, October 2006. As of April 26, 2012:
http://www.cbo.gov/publication/18187

——— "Statement of Carla Tighe Murray, Evaluating Military Compensation, before the Subcommittee on Personnel, Committee on Armed Services, United States Senate," April 28, 2010.

———, *Reducing the Deficit: Spending and Revenue Options*, March 2011. As of April 26, 2012:
http://www.cbo.gov/publication/22043

CPS—*See* Current Population Survey.

Current Population Survey, *Design and Methodology*, Technical Paper 66, Issued October 2006. As of April 26, 2012:
http://www.bls.census.gov/cps/tp/tp66.htm

Department of Defense, *Career Compensation for the Uniformed Forces: A Report and Recommendation of the Secretary of Defense by the Advisory Committee on Service Pay*, Washington, D.C., 1948.

———, *Report of the President's Commission on an All-Volunteer Force*, Washington, D.C., 1970.

———, *Report of the Ninth Quadrennial Review of Military Compensation*, Washington, D.C., 2002.

———, *Department of Defense Instruction 1145.01: Quality Distribution of Military Manpower*, Under Secretary of Defense for Personnel and Readiness, Washington, D.C., September 20, 2005.

———, *Report of the Tenth Quadrennial Review of Military Compensation*, Washington, D.C., 2006. As of April 26, 2012:
http://www.defense.gov/news/QRMCreport.pdf

———, Opening Summary–Senate Armed Services Committee (Budget Request), delivered by Secretary of Defense Leon E. Panetta, Washington, D.C., February 14, 2012. As of April 22, 2012:
http://www.defense.gov/Speeches/Speech.aspx?SpeechID=1651

Goldich, Robert L., "Military Pay and Benefits: Key Questions and Answers," Congressional Research Service Issue Brief, June 15, 2005. As of April 26, 2012:
http://govwin.com/knowledge/military-pay-and-benefits-key/16727

Grefer, James, with David Gregory and Erin Rebhan, *Military and Civilian Compensation: How Do They Compare?* Alexandria, Va.: Center for Naval Analyses, CRM D00249636.A4/1REV, August 2011.

HOM—See Bureau of Labor Statistics, *Handbook of Methods.*

Hosek, James, *A Recent History of Military Compensation Relative to Private Sector Compensation*, Chapter 2, Volume 2 of Department of Defense, *Report of the Ninth Quadrennial Review of Military Compensation*, 2002. As of April 26, 2012:
http://prhome.defense.gov/RFM/MPP/qrmc/Vol2/v2c2.pdf

Hosek, James, Jennifer Kavanagh, and Laura Miller, *How Deployments Affect Service Members,* Santa Monica, Calif.: RAND Corporation, MG-432-RC, 2006. As of April 26, 2012:
http://www.rand.org/pubs/monographs/MG432.html

Hosek, James, and Paco Martorell, *How Have Deployments During the War on Terrorism Affected Reenlistment?* Santa Monica, Calif.: RAND Corporation, MG-873-OSD, 2009. As of April 26, 2012:
http://www.rand.org/pubs/monographs/MG873.html

Hosek, James, Michael G. Mattock, Michael Schoenbaum, and Elizabeth Eiseman, *Placing a Value on the Health Care Benefit for Active-Duty Personnel*, Santa Monica, Calif.: RAND Corporation, MG-385-OSD, 2005. As of April 26, 2012:
http://www.rand.org/pubs/monographs/MG385.html

Hosek, James, and Trey Miller, *Effects of Bonuses on Active Component Reenlistment Versus Prior Service Enlistment in the Selected Reserve,* Santa Monica, Calif.: RAND Corporation, MG-1057-OSD, 2011. As of May 1, 2012:
http://www.rand.org/pubs/monographs/MG1057.html

Hosek, James, and Jennifer Sharp, *Keeping Military Pay Competitive: The Outlook for Civilian Wage Growth and Its Consequences*, Santa Monica, Calif.: RAND Corporation, IP-205-A, 2001. As of April 26, 2012:
http://www.rand.org/pubs/issue_papers/IP205.html

Hosek, James, John T. Warner, and Beth J. Asch, *An Analysis of Pay for Enlisted Personnel*, Santa Monica, Calif.: RAND Corporation, DB-344-OSD, 2001. As of April 26, 2012:
http://www.rand.org/pubs/documented_briefings/DB344.html

Kahneman, Daniel, and Amos Tversky, "Prospect Theory: An Analysis of Decision Under Risk," *Econometrica*, Vol. 47, No. 2, 1979.

Kaiser Family Foundation and Health Research & Educational Trust, *Employer Health Benefits, 2010 Annual Survey,* Menlo Park, Calif., 2010. As of May 25, 2012:
http://ehbs.kff.org/2010.html

Office of the Secretary of Defense, Military Compensation, "Basic Allowance for Housing Levels and Increases," no date. As of April 26, 2012:
http://militarypay.defense.gov/pay/bah/01_level_05.html

Sellman, W. Steven, "Predicting Readiness for Military Service: How Enlistment Standards are Established," Draft prepared for the National Assessment Governing Board, September 30, 2004.

Simon, Curtis, and John Warner, "Managing the All-Volunteer Force in a Time of War," *The Economics of Peace and Security Journal*, Vol. 2, No. 1, 2007.

U.S. Code, Title 37, Chapter 19, Section 1009, "Adjustments of Monthly Basic Pay." As of April 26, 2012:
http://www.law.cornell.edu/uscode/text/37/1009